RV Living:

A Practical Guide To The Full-Time RV Life (RV Living, RVing, Motorhome, Motor Vehicle, Mobile Home, Boondocks, Camping)

Weston Rosser

©Copyright 2017

All rights reserved. No portion of this book may be reproduced - mechanically, electronically, or by any other means, including photocopying- without the permission of the publisher.

Disclaimer

All rights reserved. No part of this publication may be reproduced, distributed, or transmitted in any form or by any means, including photocopying, recording, or other electronic or mechanical methods, without the prior written permission of the publisher, except in the case of brief quotations embodied in critical reviews and certain other noncommercial uses permitted by copyright law.

The information provided in this book is designed to provide helpful information on the subjects discussed. The author's books are only meant to provide the reader with the basics knowledge of the topic in question, without any warranties regarding whether the reader will, or will not, be able to incorporate and apply all the information provided. Although the writer will make his best effort share her insights, the topic in question is a complex one, and each person needs a different timeframe to fully incorporate new information. Neither this book, nor any of the author's books constitute a promise that the reader will learn anything within a certain timeframe.

Table of Contents

Introduction 4
Chapter 1: Choosing a Type 7
Chapter 2: Buying a New RV 26
Chapter 3: Buying a Used RV 34
Chapter 4: After Purchase 43
Chapter 5: Paramount Logistics On The Road 47
Chapter 6: Making Money On The Road 57
Chapter 7: Camps, Parks and Boondocking 65
Chapter 8: Useful Items for RV Living 75
Chapter 9: Living the RV Life 84
Conclusion 91

Introduction

Have you at any point felt attracted to the idea of "the open road"? Do you feel anxious when you remain in one place for a long time? Do you long for flying out to better places, and awakening to new encounters each day? Is it accurate to say that you are prepared for an existence change, with the capacity to go out all alone and empty your way? These sentiments are a piece of the core of living the RV lifestyle. An RV, or recreational vehicle, is the vessel for living an existence of moderation, enterprise, and opportunity.

RVs are regularly utilized for individuals searching for an escape, regardless of the possibility that it's just for an end of the week. Many people purchase RVs to continue camping trips and utilize them to travel all the more quickly amid excursion or retirement. Be that as it may, for a few people, their RV is more than a type of transportation or an impermanent getaway; it's their home.

Be that as it may, what is an RV lifestyle? Fundamentally, it's the decision to forever live in a recreational vehicle, rather than a run of the mill house or condo. It's a lifestyle that is loaded with versatility, effortlessness, evolving landscape, and closeness to nature. Individuals may live in their RV for an assortment of reasons, and there is no age, sexual orientation, or identity limitation to RV living. From a young couple planning to live more cheaply, to a resigned individual searching for some peace and calm – anybody can carry on with the free life on the road.

Inspirations for RV living incorporate the yearning to spare money, live more straightforward, travel more, expand less, and feel nearer to nature. For a few, it is essentially a profound calling. For others, it's a lifestyle alternative observed to be most reasonable for themselves and their family. Whatever the explanation behind picking this way of life, RV living accompanies many remunerating encounters, and also challenges.

This book is implied for those intrigued by finding out about living an RV lifestyle, or for RV "lifers" who are recently searching for some extra tips and traps. This book can be utilized as a manual for a start or proceed with your adventure toward living a more liberated, more self-ward life. You'll find out about the diverse sorts of RVs, and the significance of picking the correct one in light of your individual needs. You'll read about the contrasts between purchasing another RV as opposed to buying a utilized RV, and how the obtaining decisions you make will be impeding to your long haul joy for RV living.

We additionally cover necessary things to consider after you have made your purchase, and how to manage natural coordinations with regards to living without a private address. We'll give you methodologies and thoughts for how to profit on the road and also advice on the best way to live comfortably and proficiently with your RV. Finally, we'll give you a practical perspective of RV living, including the many ups and downs that accompany this lifestyle.

We trust that you will have the capacity to take away something from this book, regardless of the possibility that it's quite recently some motivation or real considerations concerning RV living. We trust that you can utilize this book as a reference for considering or effectively seeking after an RV lifestyle. We believe that this book will help you increase some understanding and inspiration for exploring after your fantasies of living a more happy, advancing life on the open road.

Chapter 1: Choosing a Type

What characterizes an RV, or recreational vehicle? An RV is a vehicle that likewise contains decent features, for example, a bed, kitchen, and restroom. The features inside the RV can change enormously contingent upon the sort, model, plan, and alterations it has.

As indicated by the Pennsylvania Recreation Vehicle and Camping Association (PRVCA), an RV is characterized as "a vehicle that consolidates transportation and temporary living quarters for travel, diversion, and camping." Try not to be off-put by "temporary" – RVs are extraordinarily assorted in their styles and comforts, and they can most unquestionably be composed or adjusted for the lasting living.

Making the Right Choice

There is a broad range of RV types, models and brands available for rent or purchase from a variety of sources. Choosing the right RV will be one of the most important decisions you will make during the process of transitioning to living on the road or at campsites and parks. The first thing to know is that there are two categories of RVs: motorhomes and trailers. A motorhome RV only means that the whole thing is motorized, while trailers (sometimes referred to as campers) need to be towed and cannot move on their own.

While it may feel like a natural choice, there are a lot of things to consider before choosing whether to buy a motorhome or a trailer. Among these two categories, there are specific types and models that you can choose from, based on your individual needs, interests, and budget.

Each type of RV has advantages as well as disadvantages, so carefully researching each one is vital before making your final decision.

Getting a Motorhome

Class A motorhomes are the biggest RVs you can get, and the most costly. Class A RVs are presumably what rings a bell when you consider huge, gorgeous motorhomes. They are considered "bus-like" and independent, alluding to their incorporation of crisp and waste water stockpiling tanks. Class A RVs are regularly portrayed as "incorporated" motorhomes because the driving territory is coordinated into the living range — making an extensive, healthy body.

Price Range: Between $50,000 and $900,000, as per RVUSA.

Average Size: 25 to 40 feet.

Pros

- Spacious.
- More space, more room for individuals, furniture, and belonging.
- Additional storage room.
- They can have "storm cellars," and there's regularly a decent measure of freight stockpiling for keeping things not implied for quick utilize.
- Expandable living quarters.
- Many Class A RVs have slide-out and overlay out abilities, which take into account more space and usefulness. This is extraordinary for an active family, or for somebody who needs space for various exercises in different circumstances.
- Additional amenities.

- This is a major one. Class A RVs have more space for extra amenities that you wouldn't be
- Ready to get in littler vehicles, for example, a washer and drier, full restroom, and space for an expansive amusement focus. They can likewise be furnished with main rooms and bigger kitchens.
- Locally available generators.
- Many Class A RVs are outfitted with generators, so you can, in any case, have control regardless of the possibility that there are no hookups accessible.
- Extraordinary for lifers.
- Class A RV's can be furnished with all that you have to live serenely, notwithstanding for long stretches of times without hookups.

Cons

- Costly.
- The purchasing price is very steep, and that is excluding the expenses of support, repairs, operation and protection.
- Scary.
- This is a LARGE vehicle that you'll be driving around, and that can be a frightening thought for some individuals. You should be twice as engaged and mindful of your surroundings contrasted with compatible drivers. Driving a Class A RV requires a considerable measure of "vehicular certainty," and additionally extensive information of road directions and behavior.
- Navigational impediments.

- There are spots that you just won't have the capacity to take your Class A RV, because of its massive size and trouble with winding, limit roads. It may not be an extraordinary decision if you are planning to do a considerable measure of going in the wild.
- Once you've set yourself up at a campground, you should experience the lengthy procedure of storing everything, regardless of the possibility that all you need to do is to make a quick excursion to some place in the close-by town.

Take-Away Message: If you have a high spending plan and you're searching for something with heaps of living and storage room, a large cluster of civilities, and space for different relatives, a Class A RV is a decent decision. You ought to likewise be a sure driver and willing to remain constant about reliable support, be that as it may. A Class RV is excellent for lifers who will put in the money and exertion of driving and upkeep, in return for more comfortable and lavish living lodging.

Class B motorhomes are littler and lower profile than Class An and C motorhomes. They are considered "van-like" and alluded to as "semi-incorporated," because of their consolidated characteristics of limited living space and taxicab zone. They are like Class C motorhomes. However, they do not have an overhead cab unit and are based on a full-sized van skeleton instead of two separate units.

Price Range: Between $37,500 and $71,000, as per RVUSA.

Average Size: 17 to 19 feet.

Pros

- Offer basic comforts.
- Incorporates resting ranges, kitchen, ventilating, and can offices.
- Easy to drive.
- Because of their littler size, Class B motorhomes are considerably less demanding to drive and more advantageous for bringing down smaller roads with sharp turns. They are additionally simpler to store away and fit in littler parking spots, giving you more outside space around your assigned stopping region in case you stay outdoors. You can take them on errands a great deal more proficiently than bigger motorhomes, making shopping and trips to town considerably simpler.
- Less expensive.
- More affordable than Class A motorhomes. They will go through less fuel and require less upkeep.

Cons

- Littler in size.
- This class has less living space accessible. They are intended to quickly suit two individuals so that Class B motorhomes wouldn't be an awesome choice for a vast family.
- Not the least expensive.
- They are as yet pricey in correlation with some other RV choices.
- Almost no space for capacity.
- You should feel comfortable downsizing your belonging to the very rudiments and necessities. You'll have to get used to utilizing littler than-typical machines.
- The absence of utilities.
- You'll, for the most part, require snare ups (interfacing with an RV stop's service administrations) for clothing and different offices that could be incorporated into greater RVs.

Take-Away Message: If you have a direct spending plan and like to appreciate essential conveniences without much debauchery, Class B motorhomes are a decent choice. They may likewise be ideal for you on the off chance that you will trade space and capacity for comfort and driving mobility. For the changeless living, this RV is best for single individuals or couples without youngsters.

Class C motorhomes are littler, both in size and price range, than Class A motorhomes. They are alluded to as "Nook" because there is an overhead segment, or taxi over, on the front of the body, for the most part, utilized for a bed or capacity. Class C makers assemble the living range independently and then connect it to a current truck or van skeleton.

Price Range: Between $42,000 and $121,000, as indicated by RVUSA.

Average Size: 20 to 30 feet.

Pros

- Adjusted size.
- It has every one of the accommodations of a bigger motorhome at a littler scale and a less high price
- (for the most part). They regularly incorporate lavatory offices, dozing quarters, and kitchen space.
- More space.
- They have more living space than Class B motorhomes.
- Capacity and overlap out choices.
- Class C motorhomes can be outfitted with crease away tables and lounge chairs to make more space for resting or different exercises.
- Valuable overhead area.
- They have an overhead compartment for additional capacity. The taxicab over the segment is extraordinary for having a disengaged resting zone or a concealed storage room.
- Less demanding handling.

- Less demanding to drive and oversee than Class A motorhomes.

Cons

- Not exactly "A grade."
- Less space and some of the timeless extravagances than Class A motorhomes.
- Still very expensive.
- Can be costly to purchase, keep up, and work. Like Class A motorhomes, the additional space and enhancements offered with Class C motorhomes liken to a higher price tag.
- Still very huge.
- May scare to drive and hard to calmly go around urban zones. Like Class A motorhomes, having an extra vehicle for errands and overnight trips is more perfect because of its substantial size and driving difficulties.

Take-away message: If you have a bigger family or need more space, however, have a constrained spending plan, picking Class C motorhomes are a decent alternative. They consolidate the additional features and courtesies of Class A motorhomes while having to some degree littler size and greater drivability, like Class B motorhomes.

Getting a Trailer

Travel Trailer

Travel trailers are non-mechanized, separable trailers that should be towed by an auto, van, or truck. They arrive in a large arrangement of sizes, running from little end of the week campers to a bigger trailer with a full washroom and different offices. They can oblige somewhere in the range of two as far as possible up to ten individuals, contingent upon the size and model.

Price Range: Between $9,000 and $76,000, according to RVUSA.

Average Size: 10 to 36 feet.

Pros

- Different in size.
- Travel trailers can arrive in an extensive variety of sizes, so you can all the more effectively pick which model is best for your necessities and spending plan.
- Towable.
- Can be towed via autos and vans and also pickup trucks. Litter sizes are incredible for simple pulling crosswise over a nation.
- More reliable and easier to keep up.
- The way that it is not mechanized implies that fewer things can turn out badly.
- Various customization.
- Travel trailers can be intended to be as straightforward or intricate as you craving.

- Separable.
- This implies you can stop it someplace while you utilize a conventional vehicle to travel, investigate, or run errands.

Cons

- Requires a suitable towing vehicle.
- You'll need an auto, van, or truck to move around your travel trailer. You'll have to feel comfortable towing a large trailer behind your vehicle.
- Unbalanced handling.
- Can be hard to move while driving, with an additional danger of tail swing and extremely constrained capacity to switch.

Take-away message: Travel trailers join the basic pleasantries and living quarters of a motorhome with the accommodation of having the capability to separate them from a typical vehicle. They are awesome on the off chance that you just have a van or auto and you're on a littler spending plan. If you like adaptability, there is an assortment of size and outline choices for travel trailers.

Fifth-wheel trailer

Requires a suitable towing vehicle. You'll need an auto, van, or truck to move around your travel trailer. You'll have to feel comfortable towing a large trailer behind your vehicle.
Unbalanced handling. Can be hard to move while driving, with an additional danger of tail swing and extremely constrained capacity to switch.

Take-away message: Travel trailers join the basic pleasantries and living quarters of a motorhome with the accommodation of having the ability to separate them from a typical vehicle. They are awesome on the off chance that you just have a van or auto and you're on a littler spending plan. If you like adaptability, there is an assortment of size and outline choices for travel trailers.

Pros

- Loads of extra space.
- With regards to looking at trailer RVs, fifth-wheel trailers have more room and flexibility as far as living quarters. They can have a lot of conveniences like Class A RVs, for example, a full kitchen, central room, spacious washroom, and even space for washer and dryers in bigger models.
- Slide-out mechanics.
- Equipped with up to three slide-out features for included space and flexibility.
- Less demanding to handle.

- Can be less demanding to tow than travel trailers because of the gooseneck augmentation that includes use on the inside.
- Differing customization.
- Trailers can be intended to be as straightforward or perplexing as you longing.
- Less demanding upkeep.
- More dependable and simpler to keep up than motorhomes.
- Separable.
- You can stop it some place while you utilize a conventional vehicle to travel, investigate, or run errands.

Cons

- Needs the overwhelming activity to go ahead.
- Requires a 1-ton pickup truck. Unlike a travel trailer, you won't have the capacity to tow fifth-wheel trailers with standard autos or vans.
- Swarmed treks.
- On longer trips, travelers won't have the capacity to remain in the towed RV amid the drive, which can make for a confined pickup truck for bigger families.

Take-away Message: On the off chance that you'd like to join space and extravagance with the comfort of distinctness, fifth-wheel trailers are a sure thing for a better than the average sized spending plan. Fifth-wheel trailers are extraordinary for lifers who need to appreciate bigger living quarters and offices, while as yet having the capacity to confine it and utilize a different vehicle for day trips.

Things to Consider Before Buying (New or Used)

If you're planning on buying a new or used RV, you want to make sure you know what you're getting into, and whether or not you're making the right decision for your individual needs and that of your family. This could be one of the biggest purchases you will ever make, so rushing into it will only slow you down and cost you more in the long run.

Here are some things to keep in mind when choosing what type of RV to invest in:

What is your realistic budget?
This may seem obvious, but your budget, savings, and income will have a large impact in determining which RV is right for you, even after the purchase has been made. If you have a high budget, it's important to consider your future income when thinking about maintenance, repair and insurance costs as well.

How much space do you need?
Be honest with yourself. If you enjoy collecting books and trinkets or you have a lot of items that you don't wish to give up, then size and storage are something you will need to factor into your budget. However, if you are happy living on bare necessities, or if you have efficiently downsized to a select few belongings, you might consider a smaller sized RV to save money.

How big is your family?
If it's just you and you're significant other, having a large RV may not be as important as other factors such as

convenience and price. However, if you have more than two people or are planning to have additions to your family in the future, space and privacy should be significant factors in the decision process.

How much off-road traveling do you want to do?
Some RVs are better for exploring uncharted territory and rough camping, while others are more suitable for the main road and RV park travels. If you enjoy driving through backwoods and cross-country, it may be more ideal to have a vehicle you can detach from your RV.

That being said, many RV sites will offer enough free beauty and recreation that you will not need to go very far from your camper to enjoy nature. See Chapter 6 for more information about camps, parks, and boondocking sites.

Do you want to be able to drive separately for quick or temporary trips?
If you just need to go across town to the grocery store, you may have issues consistently driving your large Class A motorhome to and from malls and shops. In fact, that will likely be impossible in most cases.

If you are frequently staying in more urbanized areas or cities, a smaller or completely detachable RV may be the way to go. However, if you are fine walking or biking to most places or just enjoying the areas around your campsite or park, drivability may not need to be as much of concern for you.

Where will you be living and traveling?
There is a significant distinction between residing in an RV in blanketed wild versus in a hot, dry betray. You will require an RV that is high and outlined with weatherproof features, particularly on the off chance that you know you will remain in ranges with extreme temperatures and erratic climate. For instance, a few RVs will have exceptionally outlined tops on the front, to help divert rain or wipe out effect from the cruel weather, including snow.

What are your leisure activities?
On the off chance that you like to invest the greater part of your energy outside doing exercises, for example, angling and climbing, having a littler inside space might be adequate for your necessities. Be that as it may, on the off chance that you appreciate doing makes or indoor exercises, for example, watching motion pictures and cooking, you should figure these the size and design of your RV. Make an effort not to get excessively snared on a shoddy price tag on the off chance that it implies yielding the space essential for doing the things you cherish – your long haul satisfaction is what's vital.

What amount of upkeep will it require?
Like anything that gets utilized every day, your RV will need the visit and regular maintenance. It's much better to perform general upkeep on your RV than to hold up until it will require all the more expensive repairs. All things considered, you've made an enormous speculation, so you ought to will to deal with it. For more data about building up an entire support arrange, see Chapter 4.

Consider protection.

The greater and more costly an RV is, the more protection you will likely need to pay. The kind of protection that you get will be subject to the right sort of RV you have, so it's vital to get protection that works for your one of a kind make and model.

Consider efficiency.
Mileage is characterized as "the fuel productivity connection between the separation voyaged and the measure of fuel devoured by the vehicle." Some portion of owning and living in an RV implies having a solid handle on your RV's efficiency abilities and constraints. Distinctive sorts of RVs will devour another type and measure of fuel, which can cost nearly contingent upon different elements.

To tow or not to tow?
Choosing getting a motorhome versus a trailer can be troublesome. All things considered, one of the principle questions you have to ask is regardless of whether you'd like your RV to be separable from a vehicle. If you have a reliable truck or auto and you appreciate having the capacity to do quick errands or go ahead rough terrain enterprises, towing a trailer may be better for you. In any case, if you cherished having the capacity to drive your house wherever you need and had it all together in one unit, you may get a motorhome.

Furthermore, you have the choice of towing your truck or auto with a motorhome. With vehicle towing, there is a progression of legal prerequisites and towing strategies to consider–check your nearby laws and directions to make sure you're doing it the privilege and lawful way.

Towing.

If you have towed a trailer RV, you should ensure you have the association and pulling process made sense of. Did you realize that there are indeed various sorts of vehicle wiring frameworks and towing strategies? You should consider the kind of installation that your vehicle has, figure out how to advantageously interface and disengage your trailer or vehicle, guarantee that your signal framework is working appropriately, and a great deal more.

You will likewise need to understand all the best possible parts for obligingness driving, brake control, and towing weight limit. While it might appear like a great deal to take in at, to begin with, figuring out how to pull a trailer is like figuring out how to drive an auto – it takes learning and a considerable measure of practice!

This may seem like a lot of research to perform, yet it will be justified, despite all the trouble at last. You wouldn't purchase a house or an auto on the spontaneous without preparing, and buying an RV ought to be a similar way. You may locate that living in an RV is n't exactly as incredible as you thought. Then again, you may find that it's something you were destined to do.

Eventually, you ought to utilize a blend of research and experience to settle on your choice, remembering the large assortment of RVs accessible available. Regardless of the possibility that testing one write didn't work out the way you trusted, don't get disheartened! If this is something you feel emphatically about, continue doing research and picking up as much perspective as you can.

Help yourself to remember your objectives and the purposes for your enthusiasm for purchasing an RV. Be patient and open-disapproved – make your purchase with certainty and energy, both of which are required for making your dedication to RV living!

Chapter 2: Buying a New RV

Since you've picked the ideal sort of RV suited to your particular needs, objectives and spending plan, you are entrusted with choosing whether to purchase another or utilized RV. You've experienced everything to consider in the former area, and you have an entirely quick thought of precisely what sort of RV is best for you, and what comforts you might want to have.

Presently you are confronted with purchasing a more expensive, brand new RV or buying an RV that has as of now been utilized and may have a few wrinkles to work out. Purchasing another RV can be a fantastically energizing knowledge, however, as with making any massive venture, it accompanies a few professionals and also cons. In case you're inclining toward purchasing another RV, here are a few things to consider that can help you settle on the choice of regardless of whether another RV is the approach.

Advantages of a new RV

It's new!
New things are sparkling, energizing, and extraordinary to take a gander at. You know precisely what you spent your money on, and you have set up a level of trust in your purchase since you realize that you've put resources into quality and freshness. Nobody's lived in it sometime recently, so you can make it your own.

Current features and comforts.
A new RV will have all the new devices and innovation for your happiness and diversion. It will have avant-garde particulars and usefulness, so you can feel more satisfied that it will work and drive the way it ought to, with insignificant hazard.

You have a guarantee.
On the off chance that something out of your control tragically turns out badly after your purchase, you will have the security of realizing that the maker will have the capacity to make any essential repairs or substitutions to guarantee it's in the secured condition before the deal was made.

Customization.
In case you're purchasing another RV, you can pick and pick the sorts of features you might want it to have an extensive variety of alternatives for space and extravagance.

Cons of Another RV

More costly.

With regards to paying for size and esteem, you will in all likelihood be paying more for another RV than if you somehow happened to purchase a utilized RV with a similar style and civilities. You might spend a considerable measure of money on an RV that you could get used at a less expensive cost, while not surrendering quality. This is the place you should weigh together your interests for customization, makes and models, conveniences and price range. You'll need to devote a considerable measure of time to inquire about.

While this is additionally the case with purchasing a utilized RV, buying another RV will involve perusing past many garish ads on assembling sites, and you should get down to the level down points of interest for every RV posting without offering into cushion and shimmer.

Devaluation.
When your brand-new RV leaves the parcel, it will deteriorate, or lose the estimation of its purchase price.

Research Phase

Picking a floor plan type.
The floorplan of your RV will decide the accommodation and usefulness of your day by day exercises. Having a floor plan that doesn't suit your necessities can bring about lots of worries not far off, and redesigning another floorplan can add a lot of additional costs.

While picking a floorplan, you will take a gander at the position and situation of each space, including the shower, latrine, bed, racking, kitchen, lounge chair, TVs, table space and more. Which areas are most critical to you? Which zones might you want to be more private rather than more open?

Consider how slide-out features will change the flow of the floorplan. Distinctive makes and models of RVs offer different types of floor plans, so focus on everyone that catches your advantage.

Looking into producers and perusing client surveys.

Keeping in mind the end goal to locate the best producer and model that is ideal for you, you should do a lot of research. Search for sites that are trustworthy and have finished "About Us" and "Get in touch with Us" segments. To discover a maker, you should pick a make and model. RVUSA records the accompanying makes as the top alternatives in the business:

- Timberland Waterway
- Cornerstone
- Coachmen
- Jayco
- Heartland RV
- Winnebago
- Thor Engine Mentor
- Dutchmen
- Grand Plan
- Prime Time
- Fleetwood
- CrossRoads

Make a rundown of features and comforts and number them from most vital to minimum critical. Here are a few cases of features to search for in RVs relying upon what your taste and interests may be:

Numerous capacity compartments with racking.

- Broiler and fridge size.
- Television and excitement.
- Sunlight based boards.
- Gas or electric apparatuses.
- Windows and sky were facing windows.
- Security blinds and shades.
- Dampness and temperature control.
- Dead air space and clamor control.
- Surge insurance.
- Incapacity facilities.

Have a rundown of inquiries and remarks prepared on the off chance that you will be conversing with a business agent. On the off chance that you require motivation for what to get some information about, or in case you're pondering about the genuine nature of a particular RV – read client audits! Hearing and perusing about direct encounters that other individuals have had with either make, and model is an exceptionally valuable approach to increase some practical understanding on the RV you're keen on.

Deciding your installment technique and plan.

Will you be paying everything in advance, or making incremental payments over a specific timeframe? Keep in mind that purchasing an RV is like buying an auto or house – it will to a large degree affect your credit, and the way you fund it will tail you into what's to come.

You should set up a thorough arrangement for your RV installments, which likewise needs to incorporate the cost of expenses, charges, fuel, upkeep, protection, repairs and remodels. In case you're contemplating applying for a line of credit to back your new RV, remain mindful of financing costs and blowing up prices. Devote a notepad only for your RV's funds and monitor everything about the procedure.

Where to Look When Buying a New RV

Looking for another RV can overpower. On the off chance that you do a web search "purchase another RV," you will get thousands of results from various sources offering a range of types and brands. The key is not to give yourself a chance to get totally sold on one brand or vendor without doing research on some other kind. Do extensive research and visit numerous dealers before making the purchase. Contrast prices and features with ensure you are getting an ideal arrangement.

Make a point to get a chance to physically take a gander at the fundamentally the same as or correct RV that you will purchase. Seeing it in person will give you a unique perspective than taking a gander at pictures on the web. An ideal approach to do this is to visit nearby RV dealerships. Know about pushy business people, in any case, and don't give yourself a chance to settle on any choices without feeling 100% sure.

If setting off to a physical dealership isn't a conceivable, some dealership sites will offer walkthrough recordings and itemized floor plan diagrams. Search for both physical and virtual spots that will give you clear, in advance data and just converse with merchants that will provide direct responses to any of your inquiries.

Contingent upon your interests and spending plan, leasing an RV may be a more secure wager than purchasing another one immediately. On the off chance that you've done an absolute measure of research and you're still vacillating about what RV is ideal for you, consider leasing or getting for a tryout.

Contingent upon your area, there are a few sites and businesses where you can incidentally lease an RV at an extremely reasonable price. On the off chance that you have a companion or relative who possesses an RV, consider getting it or running with them on a camping trip. These are critical approaches to experience the RV lifestyle without doing a significant monetary duty immediately. Likewise with most things, you never truly know until you attempt!

Chapter 3: Buying a Used RV

Because something has been already used doesn't imply that it's harmed or less trustworthy. Besides, you don't need to purchase a brand new RV to appreciate a high caliber, dependable motorhome or trailer. You may even have the capacity to locate a "used" RV that has scarcely ever been driven or pulled some time recently. Purchasing a used RV can spare you lots of money, and finding the right merchant can mean the contrast between making a top notch buying and getting misled.

You will presumably need to do significantly more research than if you were buying another RV. However, that is not an awful thing. Look around as much as you can and ensure you have clear rules and particulars you'd like while keeping an open personality and being willing to make bargains. With regards to regardless of whether you ought to purchase another or used RV, there is no correct answer. Scan for something that works for you, and just make the purchase on the off chance that you felt sure and guaranteed that you are settling on a sensible decision for your original goals and spending plan.

Pros of a Used RV

Less expensive. - As beforehand specified, a used RV will, for the most part, be more affordable than another RV with similar features. On the off chance that you are on a tight spending plan, purchasing a used RV may be the best decision, the length of you are cautious in the buying procedure.

Trying out RV driving. - Contingent upon the type of RV you're intrigued, you will likely need some practice on the road and additionally in parking areas. Dinging up you're officially weathered used RV 's not really as wrecking as wounding your sparkling, brand new RV.

Trying out RV living. - If you are anticipating setting something aside for two or three years before making another purchase, having a transient used RV may work well for you amid that era. You'll have the capacity to spare money to buy another and more significant RV later on, while as yet having the ability to live in an RV while you spare. This is extremely reliant on your wage, objectives, and living circumstance prerequisites.

Living in a used RV for some time can give you a decent time for testing if you are still going back and forth about putting resources into something substantially greater and better. On the off chance that you are as yet making the most of your used RV as time passes by, you may choose there is no compelling reason to purchase another one.

A gem was waiting to be discovered. - You have a shot at purchasing a beautiful RV for an unusual arrangement. Many individuals offer scarcely used RVs that have extravagance pleasantries and fresh, cutting edge additional items and adjustments. You may even discover different RVs with increases and redesigns that you wouldn't create the new market.

Cons of a Used RV

Quality and mileage. - If you anticipate doing a considerable amount of traveling, a used RV may have a higher danger of issues than another one. They will likely have more mileage, and doing a ton of traveling in an officially very much gone RV can be unsafe to the extent potential requirement for repairs.

You risk getting misled or deceived. - Lamentably, this happens once in a while, and you should find a way to dodge it by doing an absolute measure of research and being firm about quality confirmation and personal investigations.

You may need to bargain in light of your needs, needs, and spending plan. - You may need to adjust the significance of size and weight against different elements, for example, price and quality. Continuously contemplate your long haul satisfaction while being practical.

Regular pitfalls. - Now and again individuals offer their used RVs for a reason besides making a buck – there's something incorrectly. You may stall out with another person's issue, and that is never something worth being thankful for. Once more, this can't be focused on enough: Exploration and quality confirmation are imperative when purchasing a used RV!

Research Phase

Picking a floorplan type.

Like picking a floorplan for another RV, you will even now have alternatives when purchasing a used RV. Nonetheless, your alternatives might be more restricted or one of a kind relying upon your financial plan or what RV merchant you're managing. Be as practical as conceivable while thinking of you as and your family's requirements for space.

Consider living in a more kept space together for a day to get a thought of what it is like to live in a given floorplan. Do all of you capacity well together? What amount of space do you requirement for cooking as opposed to engaging versus unwinding, and so forth.? Would you incline toward the washroom to be nearer to your room or the kitchen? What zone would you like the simplest access to?

Your floorplan will decide the structure and usefulness of your regular exercises, so you ought to settle on a shrewd choice about picking the one that will be most perfect for you and your family. Unlike with purchasing another RV, you might not have refreshed online floorplan pictures accessible for the used RV you are occupied with acquiring. This is the reason it's key that you get an opportunity to physically go within it before making the purchase — so you can make certain it's truly an ideal choice for your necessities.

Inquiring about producers and perusing client surveys.
Once more, you will experience a comparable procedure as purchasing another RV, yet you should give careful consideration to assembling points of interest and client audits. You'll have to single out makers that are most trustworthy and sturdy for long haul utilize, so you can feel more comfortable purchasing a used RV knowing it's a make and model that is worked given quality and strength. Exploit point by point client audits, and note what individuals are saying in regards to upkeep, repairs and any issues that appear to be reliable with that make or model.

Checking protection cites.
A security quote is a gauge for the measure of premium, or installment you will make for a security approach with a given insurance agency. You can discover this by calling or heading off to the insurance company's site and entering in a progression of data, for example, what type of RV you have, how frequently you utilize it, and regardless of whether you use it as a living space. Diverse insurance agencies will offer notable inclusions and securities, for example, add up to misfortune substitution or crisis costs.

Picking a reliable merchant.
You ought to dependably be distrustful of your RV merchant, regardless of the possibility that they are neighborly and supportive. A reputable retailer will be entirely forthright with you toward the begin, and they will express any potential concerns immediately.

Look locally, voyage through adjacent stops and attempt to get a smart thought of the merchant's profile and aims; regardless of whether they've simply not had sufficient energy to make the most of their RV completely, or on the off chance that they're hoping to dispose of it since they can not travel anymore. Try not to be hesitant to glance around before settling on an ultimate conclusion, and don't purchase without feeling it's the best reasonable alternative for you.
  When Buying

Checking the RVs past-use history.
Obtain all the past records of the RV, including insurance documents, accident claims, maintenance records, and anything else that you can access regarding the RV's history. Choose a dealer or seller that can offer you all of the RVs original owner's manuals and documentation from when it was purchased new.

Inspecting the unit for quality assurance.
- Consider this checklist for things to look for regarding the RV's quality and condition:
- Have your seller or dealer walk you through the RV. Have them show you that everything works. Is the toilet functioning as it should? Is the kitchen in top condition? How is the water tank? What's the quality of the upholstery? Does the engine start up as it should?
- Has the RV been "painted"? If so, this means the RV has a code that will indicate whether or not the RV has been sold as salvage.
- Ask the seller to tell you when the RV was last used, how many trips it's made (and where), and what problems have occurred.

- Make sure it's clear whether or not the seller owns the RV, and how they came about it themselves.
- Ask about any leaks, tire exchanges, odor problems and any other unseen quirks or problems that you should know about.

Determining a fair buying price

Look up the RV you are interested in and compare every possible price from multiple sellers.
Consider the quality, condition, and mileage. Don't let someone sell you on one single component of the RV – you want to feel confident that the whole thing meets all of your requirements. You don't want to spend a crazy amount of money on something that doesn't live up to the standard of the price you paid.

Additionally, you also shouldn't offer too little and miss out on a chance to get an RV that you know has everything you need. If you're sure that you're in it for the long haul, you should consider increasing your budget if possible (this goes for buying new as well).

Using websites such as nadaguides.com/rv or rvvaluesonline.com can be very helpful in finding prices and values, based on precise specifications that you enter depending on what you're looking for. With NadaGuides, you will begin by submitting your zip code and choosing a manufacturer. You will then be given a selection of make and models for RVs depending on what manufacturer and type you wanted.

For example, if you chose "Wildwood by Forest River," you'll see results for several travel trailers and fifth wheels with different lengths, widths, coach design, weight, self-containment, and a number of axles and slides. You can then pick out a 2017 "M-26DDSS", a fifth wheel trailer that is 30'9" and 7,203 pounds. For this trailer, the suggested list price would $31,876, and you'll be given a list of similar RVs for sale near your zip code.

For another example, you can select "motorhome manufacturers" and choose Winnebago. You might select "M-27D-Ford", a Class C motorhome at 29'5" with a side dinette floorplan. This time you'll have the option of refining your search by features such as air conditioning and appliances.

The suggested list price for your Winnebago Class C will end up at $109,725, with a given amount for sale near your zip code. This is a nifty tool to play around with and can be extremely helpful if you already know what size, weight and other specifications you are looking for.

4. Determining your payment method/plan and making an offer.
When buying a used RV, you will have the option to barter and negotiate a fair price with the seller. Be mindful and respectful of what is fair based on the research you've done, but don't allow yourself to be bullied into a higher price.

If you can't come to terms with an offer, but you love the RV in question, keep the seller's number and contact them again in a couple of weeks to see if they're still selling. They may be more likely to lower their price if there's less competition.

Chapter 4: After Purchase

Registration, Taxes, and Insurance
You have finally bought the RV you had always wanted! Presently, you have to ensure that you have all documentation and moves made for the RV's enrollment, assessments, and insurance. Picking the correct insurance for your RV is a basic stride and, shockingly, there are some real contrasts in inclusions compared and living at a habitation. That is the reason you have to guarantee that you pick an organization that will give you full scope for your portable home.

On enlistment and assessments, you should be aware of state approaches, and choose a country to be on your driver's permit and enrollment for your RVs and different vehicles. In case will do a great deal of traveling, select an express that is sensible to your individual needs and lifestyle.

Keep in mind that a PO Box is NOT viewed as a changeless address, and you may risk accidently turning into a "resident" in a state you've been in for a long time. This may seem like a great brother, yet you ought to realize that there are a lot of alternatives accessible – see Part 5 for more data about mail and charges.

Observe that the type of insurance and duties you'll have to manage will be distinctive, contingent upon regardless of whether your RV is new or used. A used RV can be a substantial portion of the purchasing price of another RV, yet it might require significantly more support and insurance scope over the long haul. When all is said in done, if there is more mileage, liability insurance costs are higher.

Creating a Well-Developed Maintenance Plan

Scheduled maintenance.
Your RV, especially if it's a motorhome, will need to have regularly performed maintenance based on the amount of mileage and usage it gets; this can be once a year or even monthly. Your RV's generator, battery, tires, and slide-outs are all important features that will need monthly or yearly inspections to ensure they are staying in top condition.

The costs of having a mechanic or RV shop inspect your RV is cheaper than the price you'll have to pay if something goes wrong due to unmonitored issues. Use your owner's manual and your warranty to figure out the best-scheduled maintenance plan for your RV.

Preventative maintenance.
This type of support involves actively making checks and small repairs to avoid bigger fiascos down the road. There are things you can do every day, week and month that will help you stay on track with your RVs condition, including keeping everything clean and lubricated, checking the battery's water levels, a pressure of tires, surge protection and much more, depending on the class and model.

Emergency maintenance.
Emergency maintenance may be needed in situations you have little control over. Some examples include if your RV blows out a tire, is severely damaged by weather or a traffic accident or your RV suddenly breaks down without fair warning. It's important to have handy emergency tools ready for situations like this (see Chapter 8), but emergencies can be reduced with increased scheduled and preventative maintenance.

Renovation or Upgrades
You will have the chance to choose whether or not you will give your RV any remodels or overhauls. For utilized RVs, you might take a gander at a significant bigger redesign extent than for another RV. Possibly you genuinely don't care for the upholstery or window treatment of the utilized RV you just purchased, despite the fact that you got an incredible arrangement on it. Or, on the other hand perhaps your family has developed, and your needs have changed, so you feel like it's the ideal opportunity for your RV to have a few remodels or updates.

Whatever your reasons, potential redesigns and updates ought to be precisely calculated into your financial plan as right on time as could reasonably be expected. Here is a rundown of the most widely recognized redesign undertakings to consider:

Solar panel installation.
More space for a need space – for instance, more bed space on the off chance that you have children, more local

if you telecommute, all the more sitting zones on the off chance that you appreciate having visitors, et cetera.

- **Redesigning or supplanting flooring.**
- **Reupholstering furniture.**
- **Enhancing the window medications.**
- **Decorating or painting dividers.**
- **Having table with a greater table and seats.**

Your RV will be your home, potentially for a long time to come. It's vital that you have a warm, fluffy feeling each time you enter your RV and set down in your bed or cook something in your kitchen. That is the reason it's vital to put some time and cash into making it feel like your own.

Give your RV special touches, and make it a place you genuinely appreciate being by the day's end. Make yourself agreeable and set aside some opportunity to kick back and enjoy the landscape, particularly when that view has been picked by you. Savor in your capacity to proceed onward or settle down at whatever point and wherever you wish.

Chapter 5: Paramount Logistics On The Road

You've done it! You've made the jump into full-time RV living. Since you've picked your optimal new or utilized RV home, and you've dealt with all the printed material and maintenance, it's an ideal opportunity to consider coordinations while traveling out and about without a private address. You will need to do regular everyday exercises, except nontraditionally.

Regardless of the possibility that you're moving around a ton or staying put in one campground or stop for an expanded measure of time, there are a ton of "ordinary" things to consider that you might not have pondered amid the rush of fervor amid the RV arranging and obtaining process.

Mail
Getting stuff in the mail, regardless of whether it's bills, magazines, letters or bundles is vital to many individuals. You may think about how on the planet you will get your mail in case you're living in an RV. That depends on! You have a few choices to look over, contingent upon the sort of way of life you'll be living and the amount you'd jump at the chance to spend on getting mail reliably.

On the off chance that you will live or traveling close by to companions or family, you can simply have your mail conveyed through them. This implies you don't need to pay anything, yet it likewise suggests you'll be depending on them significantly to oversee and deal with your mail.

In case you're searching for mail benefits that are more expert, you can lease a case at a neighborhood post office or UPS store close where you are staying or living. There is likewise proficient mail sending administrations you can purchase – this is more often than not with a club or association that you can experience to get mail or have it sent some place when you require it.

Contingent upon the club or organization, and what particular sending administrations you'd like, costs can change. Having a "home base" to send your mail to is, for the most part, the most secure wager, particularly when managing official archives and strategies, for example, vehicle enrollment and expenses.

Banking
Some portion of the appeal of living out and about is carrying on with a more practical, parsimonious life. Be that as it may, that doesn't imply that you'll escape paying assessments, managing bills, or having investment funds arrange. Internet managing an account is a super efficient administration that individuals can exploit these days, and it's an incredible approach to be able to make buys and get to reserves while out and about.

Contingent upon how much traveling you'll be doing amongst states and nations, you'll have to discover a bank (or various banks) that will permit you to make exchanges in an assortment of areas, and also make transactions on the off chance that you have more than one record. Checks, money, and Mastercards are different choices if you want to bear your cash, yet be aware of security and today's innovation – did you realize that you can utilize your Laptop / Tablet to make buys?

Notwithstanding what sort of records or monetary techniques you want to have, you should add managing an account to your clothing rundown of arrangements, additionally remembering your financial plan, travel arranges, a level of obligation and direct store data.

Taxes

As Benjamin Franklin once said, "In this world, nothing can be said to be certain, except death and taxes." You will need to pay attention to both state and federal taxes and how they will affect you long term. There's a lot of changing policies in regards to taxes and RV living, so the best way to stay informed is to do your research and talk to a professional tax attorney that can help you sort out your tax information based on your lifestyle.

Some questions you should consider regarding taxes are:

- Will you be working within a state for a certain amount of time?
- Will you be traveling and working internationally?
- Do you need a special license to work in a particular state?
- What features, such as mileage, will you be able to write off on your taxes?

Voting

In case you're living in an RV full time, you may keep running into issues in regards to voting and how this would fill to the extent which state, district, and city you would make votes for. Once more, the state you pick to be your permanent living arrangement becomes an integral factor here. You can utilize mail sending alternatives or a relative's address – simply remember that you will vote in the interest of that particular range.

In case you're anticipating voting amid race season, you may need to get a truant vote, which can be emailed to you on the off chance that you can't be physically present at the surveys.

Internet Access on the Road

In case you're similar to most present-day individuals, you likely love to have the internet, regardless of whether it's for delight, accommodation, work or the majority of the above. For an RV lifer, having the internet can likewise mean the distinction between getting paid and not having the way to work. Fear not! There is an assortment of ways you can get it together on the web while living and traveling in your RV.

Campground or stop Wi-Fi

Many camps and stops will offer free or paid remote internet that you can interface with amid your remain. The speed and cost can differ, so if this is a primary concern, then it's critical to call or research the recreation center's site previously, to ensure they offer internet administrations and the amount you should pay, if by any means.

Remote individual hotspots.

In case you're not up to speed on specific telephone sort language, a hotspot is an access point that gives you the capacity to utilize the internet. This hotspot can be on your cell phone, your tablet or a different versatile gadget that will typically fit into your pocket.

Telecommunication / Internet.

For example, AT&T, Verizon, Sprint, and T-Mobile offer various arrangements for different measures of information. While looking for portable carriers and information brands, it's likewise vital to note what their scope is. In case will be in more provincial ranges, you need to ensure you're with a transporter who still has the scope around there.

Here are a few things to consider when you're picking the measure of information (generally as KB, MB, and GB) that you will require on an every day and month to month premise:

- How many sites do you visit?
- How regularly you check, send and get emails.
- How frequently you utilize turn-by-turn headings, for example, Google Maps.
- What number of songs, TV shows and motion pictures you watch and download using the internet.
- The amount you utilize online networking and download applications.
- How much web based gaming you do.

On the off chance that you are a specialist or you do a ton of work or business on the internet, you should have reliable internet and enough information to manage you, without going over and getting charged an incredible arrangement. The best transporter will consider your requirements, answer the greater part of your inquiries and offer you an arrangement that is intended for your individual inclinations.

Trash and Waste Disposal

Presently we're getting down to the filthy subtle elements! You've most likely given this a ton of thought, or you may have slighted it totally. Regardless of whether you're managing can waste or junk, you'll need an intensive comprehension of how to discard trash and waste effortlessly and efficiently.

While it may not appear like such a major ordeal when contrasted with different parts of RV living, waste disposal is something you'll need to deal with on an everyday premise, so it's vital to be completely arranged on the best way to efficiently organize waste and maintain a strategic distance from any chaotic, sticky circumstances.

Can waste disposal

In all actuality, we as a whole need to utilize the washroom consistently and, as an RV life, the can in your RV will get a considerable measure of activity after some time. There are extraordinary sorts of toilets, and the principle ones incorporate customary RV toilets, electric showers and treating the soil toilets. The latrine in an RV looks and capacities uniquely in contrast to ordinary toilets; they, as a rule, incorporate a foot pedal (or catches for electric), a "scent plug" or bobber and a sprayer hose.

In many trailers and motorhomes, there is a unit called a "dark water tank" where all the latrine waste is put away. This should be exhausted about each few days, contingent upon the sort of can and RV. Campgrounds and parks will more often than not have dump locales where you can discard dark water waste utilizing a large hose connected to your RV.

Finally, you can decrease your dark tank utilization by:

- Utilizing less bathroom tissue.
- Flushing rapidly.
- Just utilizing hose for extremely chaotic circumstances.
- Use open eateries or cathodes when conceivable.

Keep in mind that utilizing you RV's can and discarding its waste will be a piece of your new life – so you'll need to figure out how to live with it!

Trash disposal

Since you'll be living in a little space (when contrasted with the present day Western family), it will be a considerable measure less demanding for trash and flotsam and jetsam to amass in the RV. That is the reason it's essential to perform sanitation practices reliably, for example, taking out the trash and diminishing the measure of trash that you make in any case.

When taking out your trash or waste and dumping it in a receptacle at a campground or stop, be aware of reusing approaches, and have regard for different campers by decreasing waste however much as could be expected.

Chapter 6: Making Money On The Road

In case you're prepared to plunge into full-time RV living, you should have an arrangement for your income and everyday costs. If you have a lot of investment funds to depend on – that is incredible! In any case, many individuals hoping to end up RV lifers will require an approach to keep up an unfaltering income to pay for nourishment, toiletries, fuel, stop and mail charges, telephone bill, and individual or excitement things or exercises.

In case you're considering how will profit while living and traveling in an RV, don't stretch! There are a few ways that you will have the capacity to benefit while living in an RV, in light of your aptitudes, experience, and interests.

Income differing qualities, which alludes to the different ways you make an income, will assume a significant role in your capacity to profit making progress toward supporting yourself and your family. Having just a single wellspring of revenue can represent a higher hazard for anybody, yet income differences can be particularly essential for somebody living and working in an RV, where the change of view is a great deal more typical.

Being adaptable and various with your income implies that you will have the capacity to manage breakdowns or crises all the more rapidly, appreciate additional time traveling or remaining at higher quality parks, or substantially merely permitting you to be more arranged for life out and about. A few cases of extra work you can do besides current business include outsourcing, putting resources into resources, beginning a business or getting sovereignties.

Workamping

This word looks interesting. However, it's a mix of "work" and "outdoors" – the ideal portrayal of somebody who rests in an RV and works little maintenance or full-time while carrying on with an itinerant way of life. "Work" can allude to business or outsourcing in the joint sense. However, it can likewise mean any action that you do in return for something.

The sort of working employments that are accessible are:

Media transmission or virtual work.

Employment that you can do from home (your RV) with your portable workstation. Illustrations: Book attendant, website specialist/engineer, on the web/web-based social networking advertiser, visual craftsman, information section agent, picture taker, online teacher or understudy right hand, telemarketer, and substantially more.

Travel employments.

Remembering that you can move around more effortlessly, you'll discover a lot of fascinating occupations are accessible that can take you to better places and concede your personal encounters. Illustrations: open speaker, original surveyor, benefits expert, specialist, travel operator, deals delegate, selection representative, campground worker (have, office assistant, maintenance, finishing, sales, and so on.), waterway control, farm hand, ticket taker, conveyance individual, and a great deal more!

The salary that you get and the length of these occupations can fluctuate incredibly. Client benefit or administrative rules can begin off at the lowest pay permitted by law and upwards, contingent upon your aptitudes. For artisans, architects, software engineers and other particular abilities, you may have the capacity to make amongst $50,000 and $100,000 a year, contingent upon your capacities. For venture or customer field work while traveling, you can, for the most part, gain $50,000 a year and upwards. Huge numbers of these occupations can offer long-haul work, particularly given that a considerable measure of travel or high positions is progressively popular.

Other Temporary and Seasonal Work

Regardless of whether you're quite recently searching for something here and now or you're experimenting with new things, impermanent and season work is an incredible decision if you'd jump at the chance to extend your abilities with the adaptability of not being secured to one place. Numerous occasional occupations need to do with the organization or mechanical work.

Cases: entertainment mecca or vacation spot representative, bookkeeper, client benefit, machine administrator, janitorial, maintenance worker, pumpkin/fireworks/Christmas tree deals, teacher, cultivating, or occasional Packer.

An ideal approach to discover occasional work is to check the nearby occupation postings in the region where you'll be exploring nature – you can likewise utilize sites like Craigslist or join online bulletins for RV employments.

Online Freelance Work

On the off chance that you have a laptop, and you've made sense of an ideal approach to procuring a decent internet association while living in your RV, you might consider utilizing your aptitudes and imagination to do online independent work. In case you're outsourcing, you might be characterized as independently employed, self employed entity or a transitory worker for an assortment of various organizations in different circumstances. Outsourcing on the web is an incredible approach to procuring cash from anyplace on the planet, particularly if you appreciate setting your particular calendar and working from your RV wherever and at whatever point you please.

There are a lot of spots to discover an assortment of various sorts of independent work on the web, and more sites fly up each day. The key is to efficiently find separate occupations is to develop your profile however much as could be expected while apparently showing your abilities and experience. Having an individual blog or site is an or more on the off chance that you'd get a kick out of the chance to build up an expert online nearness.

Here are a couple of the most well-known independent sites:

Upwork.com

Presently consolidated with Elance, is an extraordinary site for interfacing consultants with organizations and agencies looking for particular ventures. Up work is allowed to join too, and incorporates a wide assortment of classifications for all experience levels, including the web and versatile programming improvement, composing, outline, information science, client administration, and substantially more.

Freelancer.com

The other option to Up work, Freelancer offers an assortment of employments for consultants taking a gander at work identified with information passage, realistic or web composition, cell phone, internet programming and showcasing, Photoshop et cetera. You will have the capacity to offer for different occupations and take part in challenges to flaunt your abilities and draw in customers.

Toptal.com

In case you're a more qualified consultant seeking work for higher profile organizations, Total is a decent decision. In the wake of experiencing a thorough screening process, you'll have entry to a lot of assets and a shot at working with well-paying customers.

Craigslist.org

You may have known about Craigslist, maybe even on your scan for an utilized RV. You can likewise search for independent work on Craigslist by perusing for neighborhood activities or employments that meet your skill set. This is an impressive approach to perceiving what sort of work you can discover close-by while remaining at various campgrounds and parks.

Regardless of what aptitudes, experience or premiums you have – there is most likely something you can learn online that works for you and furthermore wins you cash. Is it true that you are incredible at writing or composing? There are many online occupations you can apply for identified with writing, correspondences, and dialect, for example, an article author, virtual associate, publicist, transcriber, web-based social networking advertiser or online instructor.

In case you're keen on innovativeness, innovation or business, you can discover employments like visual creator, PC software engineer, web designer, expert, stock broker, offshoot advertiser, or call focus delegate. You can even begin your own business if you have an amazing thought for a venture or administration!

Do some investigating and search out employments that premium you.

To abstain from being misled, guarantee that your customer has a high appraising on their profile. Begin by being adaptable, however, increment your rates as you build up your portfolio and notoriety.

Chapter 7: Camps, Parks and Boondocking

You can't be out and about, and many RV lifers remain short or long haul at camps, parks, and boondocking locales. Parks and campgrounds are areas spread the nation over that offer parking spot and on location courtesies for individuals staying or living in RVs. A few camps and stops can be very lavish by providing an assortment of advantages, for example, free WiFi and clothing offices, while others just offer the very nuts and bolts.

The best initial step to take for remaining at a recreation center or campground is to ensure you call early, ask what offices they have that you require, and on the off chance that they have enough space for your RV. Here and there, this should effectively be possible with a snappy web seek, however notwithstanding having a short telephone call can spare you a ton of time and stress. Remaining long haul or here and now at campgrounds, parks or boondocking territories can bring numerous extraordinary and intriguing encounters, as no two are truly the same.

Staying at Campgrounds or RV Parks
Professionals of Campgrounds and RV Parks

Comforts
Many parks and campgrounds offer an assortment of conveniences, for example, satellite TV, Wi-Fi, rec centers or tennis courts, flame broil and cooking ranges, swimming pools, pantries and so forth. Some may even have a ton of fun exercises or projects, for example,

practice classes, live excitement, or rentable clubhouses for gatherings.

Encountering the different qualities and elements of better places to camp your RV can be fun and energizing, particularly on the off chance that you appreciate seeing and doing new things.

Convenience.
Camp and stop locales offer you a place to rest and recover. In the wake of setting up your RV, you will have the capacity to take a load off or concentrate on your work, while taking a break from the street.

Stunning perspectives.
Contingent upon the area, many spots offer magnificent landscape and access to parks, trails, shorelines, and that's only the tip of the iceberg.

Get your mail.
With appropriate game plans, you'll have a place to send or pick up your mail.

Meeting new individuals.
You'll get the opportunity to converse with people from varying backgrounds. You'll have the chance to share stories of your RV encounters, and trade tips and thoughts regarding RV living, with your neighbors and the general population you meet on your adventures.

Cons of Campgrounds and RV Parks

You may not coexist with your neighbors.
For instance, you may discover they are very noisy and troublesome in the wake of having as of now set up camp. Be that as it may, the excellent thing about RV living is that you can only proceed onward the following day if you'd like.

Can be difficult to fit in.
The bigger your RV, the more space you'll require. What's more, that can test to discover in specific parks or campgrounds.

Charges
You'll have to pay for your allotted space. Costs can fluctuate extraordinarily relying upon the administrations the recreation center gives; however, a standard daily rate can keep running amongst $20 and $75. Be that as it may, the expenses turn out to be substantially less expensive for the month to month lease. For instance, one month of lease for a site that incorporates full hookups can be around $300, which is considerably less expensive than most houses or flats.

Higher end stops that incorporate internet and vacation destinations can cost around $600-800 a month. Remember that the cost will likewise enormously rely on upon the area.

Distributed Space (Pitch/Site) Facilities
The first designated area offices that numerous RV parks give are sewer associations, drinking water, electric and

AC control hookups, Wi-Fi and TV connection. These fundamentals will permit you to run your can through sewer connections with abstaining from gathering dark water, attach you with a lot of water for drinking and giving, and give you power and internet. Each apportioned space, otherwise called "pitches" or "destinations," are the place you'll be setting up your RV and using the offices.

RV Park and Campground Etiquette
Your kindred RV campers will be your neighbors, regardless of whether it's for one night or a while. In this way, it's vital to practice campground manners by regarding the space of others and remaining mindful of how your activities will influence your environment.

Try not to stroll through other individuals' campgrounds without authorization. Continuously teach yourself about the principles of the site, and tail them. Continuously get after yourself and your pets; utilize trash and reuse containers instead of the ground or your fire pit. Keep commotion to a base at night hours, and attempt to arrive before the arranged time so you can get your RV set up before dim.

Sewer associations and waste disposal destinations will have an arrangement of strict rules that you'll have to take after, keeping in mind the end goal to guarantee that the waste is discarded appropriately and there is no development of outside articles at the disposal site. In case you're a smoker, ensure that you smoke in assigned zones far from different RVs. When in doubt of thumb, adhere to the idea of "treat others how you wish to be dealt with."

Boondocking

Boondocking once in a while alluded to as "primitive outdoors" or "dry outdoors," is essentially outdoors in your RV for nothing, or greatly minimal effort, in a range without hookups. This can be anyplace that is lawful for you to stop your RV, yet the greater part of boondocking is out in agricultural fields and nation streets.

Many individuals boondock their RVs to save money on cash or make tracks in the opposite direction from society for some time.

While boondocking can be brave and practical, it can likewise be very severe and possibly involved in case you're not appropriately arranged. Organize your boondocking trips by guaranteeing you have crisis supplies, a lot of sustenance, full water tanks and access to the battery or solar power.

Masters of Boondocking

Simple on the wallet.
You don't pay a thing to boondock! Inside legitimate confinements, you can locate your one of a sweet spot to stop your RV that is as private or near some place as you like. This can be anyplace that is lawful to stop — from a National Park to a mall parking garage.

Comfort.
If you have a reasonable RV, you can, in any case, live serenely while boondocking for nothing. On the off chance that you have an average estimated water tank,

solar power or generators and a lot of provisions, you have the fundamentals for satisfactory and lovely here and now living while out in the wild.

Brisk stop.
On the off chance that you've been out and about for some time and need to discover a place for a snappy overnight quit, boondocking is a simple choice. It's constantly better to create a place to camp and get some rest, instead of driving when you're drained.

Isolation.
Boondocking can give you the chance to make tracks in the opposite direction from the commotion and worry of normal daily existence. At times, the best drug is to get out in nature and appreciate all that it brings to the table, including quiet, peace, investigation and experience.

Cons of Boondocking

No significant associations.
You'll have no quick access to water, sewage, and electrical hookups. Moreover, you might not have Wi-Fi or close-by access to clothing offices. Unless you figure out how to have a flag, you likely won't have the capacity to utilize your cell phone. This implies you have a constrained measure of stores to live on without other assets.

Without anyone else's input.
In case you're far from town, help will be much slower on account of a crisis. That is the reason it's consistently imperative to be completely arranged, be cautious, and dependable convey an emergency treatment pack.

Eccentrics.
In spite of the fact that some may discover this angle energizing, many individuals have sentiments of dread or fear of the obscure with regards to boondocking. An ideal approach to defeat this is to be coherent and all around arranged for an assortment of conceivable situations.

Wellbeing
Similarly, as with typical outdoors, you ought to dependably make an agenda of precautionary security measures and take after your impulses if something feels off. You'll likely be far from a healing center and manage essential components, for example, extreme, surprising climate or natural life. It's vital to utilize sound judgment, however, and perceive the level measurable conceivable outcomes of a bear assault or a burglary.

Still, you ought to dependably be set up for high hazard circumstances via conveying pepper or bear shower, Walkie Talkies and radios, emergency treatment packs and securing your site with a caution framework or monitor pooch (regardless of the possibility that you're just utilizing a sign). To abstain from drawing in bears to your site, ALWAYS keep sustenance and refuse safely bolted up with a water/air proof holder. Be careful with regards to utilizing guns, as the laws in regards to firearms and weapons can fluctuate extensively relying upon your area.

Lawfulness
You'll be cautious about where you boondock to maintain a strategic distance from unlawful outdoors ranges. On the off chance that you know the scope and longitude of the web page you need to remain at, you can check sites, for example, Boondocking.org to look with the expectation of complimentary outdoors areas. Numerous National Parks and government lands claimed by U.S. Timberland Service or BLM (Bureau of Land Management) will offer free or ease campgrounds you can exploit.

With authority consent, you might have the capacity to camp on private property in a few conditions. Make sure to dependably act deferential and considerate – never abandon anything other than your tracks! You ought to likewise abstain from encouraging untamed life, as different sorts of handled sustenances can be unsafe for wild creatures – appreciate looks from a protected separation and nothing more.

What Makes an Area Good to boondocking?

The range is useful for boondocking on the off chance that you have simple access to the driver's seat, heaps of space and permeability. The field ought to be anything but difficult to enter and leave if need be. Keep away from particularly dangerous, rough terrain trails that could harm your RV or stall out. Frequently on government lands, you'll have the capacity to discover open locales that have already been utilized by different campers. Try not to camp excessively adjacent streets or water – this can influence your level of protection and your effect on natural life.

On the off chance that you'll be building a pit fire, pick a site sufficiently far from tree limbs, ideally a site with a fire ring as of now set up. Climate and atmosphere will likewise be considered, including adjusting a level of shade accessible and ensuring you're not in peril of having a tree fall on your RV.

Power

There are a few ways that you can protect your energy use while boondocking. Continuously keep lights off at whatever point conceivable, and limit the utilization of machines and warmers. Ensure your water heating appliance and generators are off when not being used. Suiting up with a thick, warm coat and a lot of covers will be adequate for keeping warm much of the time. Introducing solar panels on your RV is an active and eco-accommodating method for using the sun to give your RV control.

Water supplies

There's an old spouses' story that boondocks who don't safeguard enough water should begin reusing dark water for drinking water – on the off chance that anything, it's awesome inspiration for your outdoors sidekicks to save more assets! Washing up and conveying. However, many holders of water as could be allowed will help moderate your water levels.

Chapter 8: Useful Items for RV Living

Since you've settled in at a camp, stop or boondocking site, and you've landed your coordinations, and position status made sense of, it's an ideal opportunity to make an agenda of all the essential and helpful things you'll require for RV Living. In case you're mostly cutting back, you ought to make a plan of the things you need in the first place, and after that moreover make a little list of "need" things.

Be incredibly saving in what you bring into your RV since space and capacity are constrained. Regardless of the possibility that you do have plenty of compartments and hole to stick things in, it's vital to remember the things that you will conceivably amass later on, and consider how having a full RV will affect your nature of living.

Wellbeing Equipment

Smart vests
These will be helpful for living out and about or being out in the forested areas or in dull territories where you should be seen from a reasonable separation.

Life vests
In case you stay outdoors alongside a lake, or you do a considerable measure of swimming or angling, you'll need an existence vest convenient to keep you or your family safe amid water outings.

Emergency treatment pack

An effective emergency treatment unit ought to be incorporated into each RV! Especially if you are boondocking, you should have a medical aid group that contains all the fundamental things to use in crisis circumstances – these things can incorporate things like an electric cover, water pocket, shriek, compass, wraps, sterile wipes, without latex gloves, and a great deal more. Two or three great brands that offer select emergency treatment units incorporate American Red Cross and REI.

Fire Extinguisher.
Each home or RV needs one if there should arise an occurrence of fire!

Spotlights and headlamps
Pick high caliber, reliable electric lamps, and lights with long-haul battery life. You can discover tried and true spotlights and headlamps at spots like REI and DICK's Sporting Goods.

Repair Gear

Toolbox.
No RV is finished without an active unit of repair tools.

Pipe tape, electrical tape, and silicone protect tape.
These things will prove to be useful for snappy repairs and maintenance.

Work gloves.
If you are reviewing or making repairs to your RV or vehicle, you'll require a full match of work gloves to protect your hands.

Dispensable gloves.
These can be lifelines in case you're managing waste disposal and other chaotic circumstances.

Little Hatchet and scoop.
These are necessary tools for setting up your campground and burrowing cathodes for waste internment – collapsible scoops are suggested for simple stockpiling.

Cordless penetrate.
Drills might be expected to mount something or send stabilizer jacks.

Electrical, new parts (wires, wire connectors and so on.).
Better to have them and not require them than to need them and not have them.

Additional ropes and hoses.
Cables and tubes can end up noticeably worn other time, and on the off chance that they get punctured without substitution, you'll likely be managing some costly issues.

Devices and Gizmos

Remote climate station.

This is a valuable device that will help you screen rain, wind, moistness, temperature and barometric weight so you can simply be set up amid your outside journeys.

Laptop / Tablet and Wi-Fi flag supporters.
For both individual and work utilize, mobile phones can be useful gadgets that you may depend on while you're out outdoors, regardless of whether that is to make a crisis call or keep in contact with your campmates. Having Wi-Fi flag promoters can encourage convey you access to information in more remote regions.

Tire weight screen framework (TPMS).
TPMS gadgets can help you check the pressure of your tires and be cautioned about dangerous collapse levels. A busted tire is unpleasant; better to be as careful as possible!

Control inverter.
You'll require an inverter to keep your RVs battery from depleting amid the utilization of power.

Walkie Talkies.
In case you're boondocking or out of energy, these can be a lifeline. Superior, long distance Walkie Talkies are undoubtedly justified even despite the cost.

Tablet or portable PC.
You'll in all probability need a dependable tablet or portable workstation to travel with you, particularly in case you're depending on utilizing it for work. Asus and Windows are two or three great brands to consider.

A Little vacuum cleaner.
For ordinary chaotic heaps, a small handheld vacuum will help you clean up around your RV while sparing space.

Solar charger.
These are extremely helpful chargers that can be fueled by daylight and give you vitality when all else is spent.

GPS made for RVs.
For traveling and achieving goals, having a GPS for your RV can have a significant effect. Garmin is a top-quality brand to consider.

Versatile Wi-Fi hotspot.
For more data, see Chapter 5 about utilizing cell phones for the internet.

12v fan.
A decent box fan with a rain sensor and internal indoor regulator can help direct air in your RV through the vent openings.

Outside and Activities

Bug Spray.

Some portion of RV living is investing lots of energy outside, so outfit yourself with great, powerful bug shower.

Bear Spray.
On the off chance that you'll be doing a considerable measure of climbing in the wild or remaining in ranges close bears, they can progress toward becoming pulled into the resemble rubbish or nourishment. Bear Splash is an incredible nonlethal contrasting option to a firearm that can in a split second discourage a bear.

Bikes.
Having a bicycle will be exceedingly helpful if you don't have a vehicle to make speedy excursions into town.

Knapsacks.
For each trek far from the RV, you ought to convey a well-made pack with you to store water, bug splash, sustenance, walkie talkies and crisis supplies.

Ring hurl and additionally badminton set.
These are a couple of simples to store gaming choices that can likewise offer you and your family hours of open air excitement.

Pile of cards as well as prepackaged games.
Card and table games are extraordinary approaches to take a break without utilizing power if the climate outside is unsavory.

Mp3 player and accomplished speaker.
If you cherish music, having an MP3 player and speaker that is sturdy and waterproof is your most logical option.

Outdoors seats and lofts.
Places and lofts are extraordinary for getting a charge out of unwinding time outside. Pick materials that are solid, yet agreeable and effortlessly put away.

Fishing gear.
On the off chance that you want to angle, pick a light-weight, top-notch angling rod shaft and a fishing supply bag with adequate supplies.

Cooking and Dining
Open air or small flame broil.
Flame broiling outside is an unusual approach to cook while outdoors, so having a compact or collapsible barbecue for your RV is an absolute necessity. Weber or Coleman brands are usually utilized for RVs.

Waterproof lighter.

You'll require a waterproof lighter with adequate measures of liquid for beginning your barbecue or pit fire.

Multi use pot.
Pick one that is multi-useful, so you can undoubtedly cook one-pot suppers without much exertion.

Blender.
A little, powerful blender, for example, a Ninja Bullet is extraordinary for any RV kitchen.

Compact espresso maker.
Pick an espresso machine that is little and effortlessly put away when not being used, to spare counter space.

Convenient ice producer.
Keep in mind that your cooler space is constrained, so an ice manufacturer can prove to be useful.

Solid metal cookware.
Your cast press skillets and other cookware ought to be prepared to hold up against wear and tear; Camp Chef is a brand to consider.

Cooking tools and utensils.
- Keep it basic; one great arrangement of blades, forks, and spoons ought to suffice.
- Serving plate.
- You're eating space likely restricted, so the serving plate can be helpful for eating, particularly if you have visitors over.
- The kind of things you have in your RV will be subject to your way of life and interests. Continue garments to a base; having one outfit for each season is a decent broad standard. Consider utilizing a Kindle or tablet as opposed to gathering excessively numerous books, to spare some valuable space. Offer need to the things you require and don't give your RV a chance to pick with stuff that you'll never utilize.

Chapter 9: Living the RV Life

You're at last prepared to begin living your gutsy and free-living RV dreams! You've picked the best kind of RV for your requirements, earnings and spending plan, you have it paid for and safeguarded, you're prepared to manage accounts and coordinations while out and about, you've picked your living and traveling arranges, and you have every one of the things you have to start your RV life.

Congrats! On the off chance that you've made it to this progression, you realize that the street to your RV life hasn't been dull, shabby or quick. In any case, there are still a few things to consider about RV living that will affect you for whatever length of time that you carry on with the way of life.

What Will the Lifestyle Cost over the long haul?

This will shift enormously, contingent upon the sort of RV you have and the type of way of life you lead. As you've presumably acknowledged at this point, living in an RV accompanies a rundown of various costs, even after the buying stage. Be that as it may, long haul RV living can be exceptionally moderate and reasonable for individuals on an extremely strict spending plan.

On a month-to-month premise, you could pay somewhere in the range of $500 to $2000; it relies on upon your costs for sustenance, protection, vitality utilize, repairs, swelling, and considerably more. By and large, RV living is a remarkable efficient decision that many individuals seek after for living all the more uninhibitedly and inexpensively.

Shouldn't something is said about Friends and Family?

Being far from your loved ones can be hard. The magnificence of living in an RV is that you can travel the nation over and visit your friends and family, without stressing over requiring somebody to house-sit for you. Be that as it may, RV living may likewise imply that you will be a long separation far from them, with diminished capacities to look after correspondence (particularly if you are boondocking).

Keep in contact at whatever point you can, and disclose your way of life to your friends and family so they know you might be distant amid particular circumstances. RV living isn't generally about being confined; indeed, you can make handfuls or even several new companions at the parks or campgrounds you visit!

How Safe is it to Live Full-time in an RV?

There are a few things, to the extent wellbeing goes, that you should consider. While out and about, you risk having things stolen or becoming involved with circumstances with other individuals, creatures or occasions that may represent a danger to you. For whatever length of time that you are dependable legitimately arranged by having a flexible arrangement, supplies and techniques for obstacles (signs, cautions, puppies, and so forth.), living full-time in an RV can be as sheltered as living anyplace else.

Professionals of RV Living

A feeling of flexibility.

On the off chance that you feel a solid draw to carry on a more daring and versatile life, RV living could be something that will bring you satisfaction and happiness. Being gotten up to speed in an everyday existence with a stationary house can infrequently make individuals feel caught. You'll never need to consider that path with your RV.

Shabby living.

When you include everything up over the long haul, RV living can be a more economical way of life than acquiring or notwithstanding leasing a house. You are expending less, which implies you have fewer bills to stress over. You will be urged to spare cash as opposed to looking for more stuff since you truly don't have the space for things that aren't fundamental. When we live in houses, we can wind up aggregating a considerable measure of stuff that does nothing for us except cause a distinction of what's essential in life.

Straightforwardness.

If you pitch your home and scale down to an RV, you will do a noteworthy wash down of your belonging. Having lots of stuff can stop up your feeling of need, and carrying on with a more straightforward, more moderate way of life can help you concentrate more on the nonmaterial parts of your life, for example, family, travel, and encounters.

Adaptability.

Has your occupation or interests all of a sudden changed? Have you chosen you to need to move to the other town, state, or nation? Would you like to switch things up and make a few redesigns to your living space? You are currently ready to do these things, without being secured to a permanent location. Moving or looking for another employment after an old one closures are as simple as beginning up your RV and taking off!

A redeeming quality.

Neediness is a developing worry in many territories of the U.S., and also over the globe. Numerous Individuals are compelled to offer their homes and live in their autos or remain with the family. RV living is an option way of life that can permit you to live on a financial plan, while as yet having your one of a kind space with the flexibility to do however you see fit, still appreciate many elements and civilities in your regular day to day existence.

Cons of RV Living

The absence of space.

On the off chance that you long for having an ample room and a full-sized kitchen, RV living may not be for you. On the off chance that you appreciate gathering things, living in an RV may make it harder to store and compose the majority of your belonging, just because of an absence of space. In case you're transitioning from a house filled to the overflow with stuff to living in an RV, you should begin making gifts and yard deal boxes.

You'll need to scale back permanently, and you may need to surrender furniture or things that have the enthusiastic incentive to you. In any case, clearing out your belonging may not be a terrible thing. Having less mess around you will probably help you remain more sorted out, and you'll be more energetic about delight from encounters, as opposed to belonging.

The absence of protection for families.

If you are living with another person, remaining in an RV implies you will all need to like each other. A considerable measure. You'll be living around other people, and you'll likely be not able to have quite your very own bit space. That being stated, the magnificence of RV living is that you will probably be stopped close to a zone with heaps of space, for example, a recreation center or backwoods. In case you're willing to get innovative and investigate a bit, you might have the capacity to discover a lot of security around your campground of a decision.

Required maintenance.

There are sure things you'll need to stress over with your RV that you wouldn't need to stress over with a house, for example, ensuring it has enough fuel and the requirement for physically discarding waste. These, be that as it may, are only a few little errands in correlation with the heap of remunerating encounters that RV living can offer.

With everything taken into account, on the off chance that you ever feel the tingle to go ahead, or only to encounter a change of view or organization, you can essentially pack up and go! Your choices are unending, and there will be nothing secures you to one range on the guide without your assent.

RV living is the ideal open door for a creature and outside mates to get genuine, honest to goodness taste of the crude excellence and energy of nature. You'll find the opportunity to witness different looks into a more primitive, characteristic condition of the world that is once in a while experienced by a large number of people in present day society.

Each morning, you can wake up to something new, energizing and enjoyable. One day, it may be an old timberland or lake in the mountains; one more day it may be a bright betray desert spring or a warm, sunny beachfront. You can go wherever you satisfy; you can travel the nation on the off chance that you wish! The street's the farthest point, and that being said – don't fear to stay out of the way and investigating new places.

Conclusion

Is the RV life calling to you?

Here are a few things to repeat before starting your exploration and making your massive buy to push ahead into your new existence of opportunity:

Pick an RV that is best for your requirements. On the off chance that you need extravagance and space yet you wouldn't fret driving a scaring motorhome, Class A may be best for you. If you are searching for something little, comfortable and moderately simple to operate, pick Class B. Do you need something that is a tad bit greater yet at the same time not exactly as costly as a Class A? Consider Class C.

On the off chance that you'd like something you can tow as opposed to drive, you can purchase a travel trailer or a fifth-wheel altered for your requirements. Besides, you'll need to measure the advantages and pitfalls of buying another or utilized RV before settling on a correct choice to buy.

Make RV living useful and conservative. To guarantee your new RV is ideal for full time living, you'll need to have an all around created maintenance plan, protection, and ultimately useful floorplans or redesigns that suit your method for living. You'll have to consider habitual coordinations, for example, getting mail, keeping money and internet utilizing techniques that work for your financial plan and inclinations.

There are several working, occasional and independent occupations that you can seek after, so profit out and about, and pick something that interests you or requires your abilities.

Choosing where to camp your RV and guaranteeing you have admittance to every one of the offices you need will assume a critical part in your new RV way of life. Moreover, you'll need to ensure you have all the best possible security hardware; repair equips, contraptions, outside things, and cooking supplies you'll have to live quickly and efficiently. Advise yourself that the more arrangements you make, the more minutes you should appreciate the opportunity of RV living.

Try not to get debilitated about the many difficulties or deterrents related with RV living. On the off chance that you feel energetic about making an RV your home and living openly and autonomously, never let anything prevent you from accomplishing your fantasies.

If you take the way of experience, travel and unlimited open doors, the RV life is sitting tight for you. May you have a brilliant voyage!

www.ingramcontent.com/pod-product-compliance
Lightning Source LLC
Chambersburg PA
CBHW070105210526
45170CB00013B/754